The Railw

and Britain's Nuclear Industry

DAVID McALONE

THE RAILWAYS AND INDUSTRY SERIES, VOLUME 1

Note on the text

The information in this book has been checked via corporate newsletters, house magazines or journals and websites of relevant organisations: The Nuclear Decommissioning Authority; International Nuclear Services Ltd.; Pacific Nuclear Transport; The Low Level Waste Repository; British Nuclear Fuels Ltd. (BNFL); Sellafield Ltd.; Magnox Ltd.; DRS and the Civil Nuclear Constabulary. In addition, various stakeholder groups and local authorities publish their minutes online.

Acknowledgments

I would like to thank British Nuclear Fuels Ltd. (BNFL) and Direct Rail Services (DRS) for allowing me access to their facilities, to take photographs and travel on a DRS freight train in 1996, and for access to DRS Kingmoor Depot in the 2000s. Thanks to Freightmaster (https://freightmaster.net) for train reporting data. Also – at the risk of developing an Oscar acceptance speech – I owe a debt of gratitude to individuals, most of them former or existing railwaymen, and I must mention a few by name: Mike Benyon and Gordon Ogden, who kept me right in the driving cab all those years ago, and Gordon also for safely conducting me around depots. In addition, Bill Wright, John Waddington, Adrian Nicholls and Gordon Edgar, who kindly allowed me the use of their pictures, and my appreciation, too, for the railway expertise of Gordon Ogden, David Ratcliffe, Adrian Nicholls and Gordon Edgar.

Published by Key Books
An imprint of Key Publishing Ltd
PO Box 100
Stamford
Lincs PE19 1XQ

www.keypublishing.com

The right of David McAlone to be identified as the author of this book has been asserted in accordance with the Copyright, Designs and Patents Act 1988 Sections 77 and 78.

ISBN 978 1 913295 65 3

Typeset by Aura Technology and Software Services, India.

Contents

Chapter 1

The Nuclear Industry

In 1956 Queen Elizabeth II opened Calder Hall power station, the world's first commercial nuclear power station. Calder Hall was not the start of Great Britain's nuclear industry and it was not just a commercial power station either, it also produced plutonium for the UK nuclear weapons programme.

The 1945 post-Second World War Labour government had a lot on its plate. It began the process of nationalising the nation's key industries, including – of course – the railways. It was planning a massive house-building programme, introducing welfare and pensions reforms and establishing the National Health Service. All this was taking place against a background of eye-watering war debt, the beginnings of the Cold War and pressure from the nations of the British Empire for self-determination. How and why it came to the decision also to pursue a nuclear weapons programme is somewhat puzzling, in particular because the United States of America refused to share its nuclear secrets with a Labour government, despite the major contribution of British scientists to the Manhattan Project. However, that is what Clement Attlee's government did, establishing the Atomic Energy Research Establishment (AERE) at Harwell, which had a reactor up and running by 1947.

It turned out that mistrust of the UK by the USA was entirely justified – not due to its jaundiced view of a socialist government, but to the fact that the British Secret Intelligence service was riddled with Britons who were passing secrets to the Soviet Union. Guy Burgess alone passed over 350 top secret documents to the soviets in 1945. The UK was on its own and consequently backed the wrong horse with gas-cooled, graphite-moderated reactors while the USA and ultimately the rest of the world developed water-cooled and water-moderated systems.

By 1950 construction of the first of the two Windscale reactors, known as 'piles', was complete. In 1953, concurrent with the operation and development of the Windscale piles, construction began on the adjacent Calder Hall reactor site. The purpose of the Windscale piles was to produce atomic weapons material. To keep the piles from overheating they were cooled by blowing air through channels in the reactor cores with the resultant hot air simply blasted up the chimney. The potential for this hot gas to generate power was not a consideration during the race to get Britain its atomic bomb but, across the river at the new Calder Hall site, power generation was the main goal. Instead of blowing air through the cores, CO2 was used as the coolant and this was passed through heat exchangers in a sealed circuit rather than simply being blown out to the atmosphere. By this time the AERE, along with other agencies such as the Atomic Weapons Research Establishment (AWRE) at Aldermaston, had been incorporated into the government's single, umbrella organisation, the United Kingdom Atomic Energy Authority (UKAEA).

In 1957 Pile One suffered its infamous catastrophic fire resulting in both piles being shut down, leaving Calder Hall to carry out both civil nuclear electricity generation and provide material for nuclear weapons. Calder Hall was designed to operate for twenty years, but in fact operated successfully for forty-seven years, closing in 2003. Neither Calder Hall nor its sister plant, at Chapelcross, were connected to the mainline rail network, but all other reactors in the UK were either connected to the rail network or moved their flasks on road transporters to a railhead.

The Calder Hall design was known as a 'magnox' reactor, an acronym derived from the cladding used to encase the uranium metal fuel rods (**mag**nesium alloy no**n-ox**idising). By modern standards they had a relatively low output, but twenty-six of them were built in the UK at eleven sites with the first ones, Bradley and Berkeley, coming onstream in 1962. They were a relatively successful design and it was hoped that they could provide a significant export boost, but take-up was disappointing with only two nations utilising the design. Although only one reactor at Latina, Italy, and one at Tokai Mura, Japan, were built, they sent their spent nuclear fuel (SNF) for reprocessing to Sellafield. In December 2015 the last of the Magnox stations – at Wylfa in Anglesey – closed down.

The second generation of reactors in the UK were known as Advanced Gas-cooled Reactors (AGRs). The key advancement of these reactors was in the type of fuel they used, a ceramic form of enriched uranium, clad in stainless steel, which allowed the reactors to operate at higher temperatures. A small prototype AGR was operated at Sellafield but was not served by rail; fifteen AGRs were built and operated on six sites. These used the same principles as the Magnox reactors, using gas to transfer the heat and a graphite moderator (a moderator was essential in almost all early reactors, slowing down neutrons to increase the likelihood of absorption by the uranium). The AGRs were relatively successful but were outperformed by various designs of water-cooled and water-moderated reactors around the world: principles that would have to be adopted by the third generation of UK reactors.

The UK's third generation of nuclear reactors came on-stream in 1995, at Sizewell in Suffolk, the site of one of the first-generation Magnox sites. Sizewell B is a pressurised water reactor using principally American and French technologies. It would store its SNF on-site for eventual disposal at the national repository, the site of which had not been identified by the time the reactor was running.

Uranium is 'bombarded' by neutrons in a reactor causing the atoms to split apart in a process known as fission. Each fission event gives off more neutrons, thus perpetuating the chain reaction. The heat generated by these events is used to produce steam which in turn drives a turbine to generate electricity. However, if you want to harvest the plutonium from this process, it (along with the 'unburnt' uranium and a highly radioactive waste by-product) must be separated from the reactor by removing what is now highly radioactive fuel. It seemed obvious to locate the plant that would separate those three components close to the place where the first reactors were already operating, so Sellafield was chosen as the site for the chemical separation plant. Nuclear fuel spends several years in a reactor; theoretically, it could run for decades but the build-up of fission products within the fuel inhibits the neutron flow in the reactor, so it is replaced every three to five years. If the fuel is to be used to make nuclear weapons it has a much shorter life in the reactor and is known as low burn-up fuel. As more Magnox sites came on-stream in the 1960s and the reactors were refuelled there was a need to move the SNF to Sellafield for reprocessing, and the nuclear industry turned to Britain's railways to do this.

Sellafield had a long and chequered history, and in the 1960s suffered from the same problems that the pile chimneys had endured in the decade before: no helpful allies and the 'do or die', 'get on with it' mentality that held sway at the height of the Cold War. The nuclear industry eventually developed a culture in which risk was properly considered and managed. The concept of 'defence in depth' would be built into plant design, enhanced with international peer review and the involvement of regulators. For example, process cells, storage ponds and waste tanks would be designed with multiple layers of containment, so if the primary containment is breached equipment is in place to extract and return material from between the layers. It became the norm for plant to be built to withstand collisions, earthquakes and sabotage, but none of this had been considered when the race was on to bring Britain her first atomic detonation in 1952. The original storage ponds (designed to cool the SNF and remove its protective alloy sheathing), the separation plant and waste silos were conceived and operated at the same time as the Windscale piles, with the same purpose: to extract plutonium for weapons. A second generation of similar plants followed them in order to service the new Magnox power stations, coming on-line with the Magnox reprocessing plant starting operations in 1964. The production of weapons-grade plutonium also continued, and the UK joined the hydrogen bomb club when it exploded a thermonuclear device in 1957.

Britain's railways played virtually no part in the nation's nuclear weapons programme, but they held a key role in the movement of SNF. Once the SNF arrives at Sellafield it is cooled in a pond before its outer alloy sheath or 'can' is removed by a process known as decanning, leaving a bare bar of uranium metal.

The separation process involves dissolving the decanned fuel rod in nitric acid, a chemical that was brought to Sellafield by rail right up until 2013. The resultant liquor passes through a series of processes to extract various combinations of elements and wastes until plutonium and uranium liquors reach the purity required. The separated, concentrated waste is stored as a liquor and ultimately turned into a form of glass (vitrified). The plutonium is further processed, depending on its proposed use, and the uranium might be mixed with plutonium for nuclear fuel or sent for enrichment.

BNFL took over the nuclear industry in 1971. By the mid-1970s the rise in popularity of oxide fuel in reactors across the world prompted BNFL to build a reprocessing plant for that type of fuel. Following the 1977 Windscale enquiry the government approved the construction of a **Th**ermal **O**xide **R**eprocessing **P**lant (THORP).

The plant was a success in that it was a huge earner of foreign currency, equivalent to some £9 billion. However, its full capacity was not fulfilled, mainly due to a change in attitude towards reprocessing. It became cheaper and safer to store spent nuclear fuel instead of reprocessing it and (due to improvements in extraction and processing) new uranium became less expensive. In addition, both THORP and its associated plants were dogged by technical problems, including a very serious leak of process liquor causing the process to fall behind. A scandal involving some workers falsifying data during the production of mixed oxide fuel was also a factor in customers losing confidence in Sellafield Ltd. THORP stopped reprocessing in 2018 and the Magnox plant was also to close in 2020. However, flask trains will run for the foreseeable future as the existing ponds at Sellafield continue to fulfil their storage function for SNF.

UK Nuclear Power Stations and Their Railheads

Power station	Reactor type	Access point to the rail network
Berkeley	Magnox	Sharpness Docks branch
Bradwell	Magnox	Southminster
Calder Hall	Magnox	No connection to the rail network
Chapelcross	Magnox	No connection to the rail network
Dounreay	FBR*	Thurso branch (siding)
Dungeness A	Magnox	Dungeness branch
Dungeness B	AGR	Dungeness branch
Hartlepool	AGR	Seaton-on-Tees branch
Heysham 1 and 2	AGR	Heysham branch (siding)
Hinckley Point A	Magnox	Bridgewater
Hinckley Point B	AGR	Bridgewater
Hunterston A	Magnox	Fairlie High (Siding)
Hunterston B	AGR	Hunterston Low Level (Causeway Siding)
Oldbury	Magnox	Sharpness Docks branch
Sizewell A	Magnox	Sizewell branch
Sizewell B	PWR	Sizewell branch
Torness	AGR	Oxwellmains ECML (Torness siding)
Trawsfynydd	Magnox	Blaneau Ffestiniog
Wylfa	Magnox	Valley
Winfrith	SGHWR*	Wool (Winfrith siding)
Windscale (Sellafield)	AGR*	No connection to the national rail network

*Additional reactors on site

Abbreviations: AGR, advanced gas-cooled reactor; FBR, fast breeder reactor; PWR, pressurised water reactor; SGHWR, steam-generating heavy water reactor.

Uranium was the major fuel for the world's reactors, though most of them need an enriched version to sustain the chain reaction. All the atoms in natural uranium (chemical symbol U) have 92 positive particles, called protons, in their centre (or nucleus). Most uranium atoms also contain 146 uncharged particles, called neutrons, in their nucleus. The sum of these protons and neutrons (or 'nucleons') is 238, so this most common form of uranium is known as U238.

Every atom of an element has the same number of protons: uranium has 92, carbon has 6, iron has 26, and so on. This number defines the element and never changes (if it does change it is no longer that element). Now, many elements have slightly different versions of themselves, a sort of *doppelgänger* that has the right number of protons to define it but has a different number of neutrons; these versions are called isotopes. One isotope of uranium is U235, which is fissile, meaning it can undergo fission or can split to form different elements after absorbing a neutron. Natural uranium in pure form may be used in a reactor, but where there are imperfections, for example neutron-absorbing materials in the fuel assembly or reactor core, it is necessary to enrich natural uranium by increasing the percentage of the fissile U235. There are several ways to do this but, in the UK, a large gaseous diffusion plant (and later a gas centrifuge plant) was built at Capenhurst in Cheshire.

The uranium at Capenhurst was received from the nuclear fuel manufacturer at Springfields, Lancashire, a large complex with a rail connection off the Preston–Blackpool mainline at Salwick. Uranium compounds and the fuel elements manufactured at the site, although radioactive, were of relatively low risk so did not require heavy shielding like SNF does and were mainly moved by road transport. The facility did use its rail connection in the twentieth century for the delivery of significant quantities of chemicals that it needed, such as nitric acid and caustic soda, but that rail traffic declined in line with national wagonload trends. The sidings fell into disuse and the loop was removed as part of the Preston–Blackpool electrification upgrade in 2017.

Low-level waste (LLW) from nuclear sites across Great Britain is stored at the Low Level Waste Repository (LLWR) at Drigg, three and a half miles down the line from Sellafield; but more about this in Chapter 5.

In 2005, as decommissioning started to become the dominant factor at Sellafield, and at the Magnox reactors, the Dounreay site in Caithness and the plethora of old research and development sites in southern England, the Nuclear Decommissioning Authority (NDA) was formed. The NDA was given responsibility for the whole industry, including Sellafield Ltd., the LLWR and DRS. The remaining operational PWR and AGR reactors supplying power to the grid were taken on by EDF, a British subsidiary of Électricité de France.

In 2008 the NDA appointed a global consortium made up of the American corporation, URS, the French Nuclear company, AREVA, and energy and construction company, AMEC, to oversee the Sellafield project. The consortium was called the Nuclear Management Partners (NMP) and it lasted until 2016 when the NDA decided that it could do the job better itself. The Sellafield workforce and local community have long been sceptical about what is seen as the constant reorganisation of the nuclear industry in general, and Sellafield in particular. NMP drew criticism for appearing to take out more than it put in but, to maintain a balanced view, it is difficult to measure what it actually brought in and the consortium did invest over £25 million into the local economy from its profits.

Chapter 2

Transport of Spent Nuclear Fuel: The Home Boards

Nuclear power stations in England and Wales were government-owned and run by the Central Electricity Generating Board (CEGB) and, north of the border, by the South of Scotland Electricity Board (SSEB). In the nuclear industry they were collectively known as the 'home boards' and had been organised that way since the 1950s. By the 1990s both the nuclear power industry and the railways were structurally changed – still government-owned but organised to prepare for privatisation. The home boards CEGB and SSEB became Nuclear Electric and Scottish Nuclear, respectively; the railways went through a plethora of organisations both in preparation for and after privatisation during the mid 1990s.

According to the Railway and Canal Historical Society's *Chronology of Modern Transport in the British Isles 1945–2015*, British Rail (BR) began to handle nuclear waste traffic from approximately 1958, although details of this are few. Several sources quote existing bogie well wagons, 'rectank' (former tank transporters) or 'weltrol' designs being used to transport them. What is known is that wagons specifically designed to carry the SNF from power stations to Sellafield emerged in 1960, from Shildon.

The design and construction of the shielded containers or 'flasks' to carry the fuel was in progress at about the same time, but it must have been unclear as to what these might weigh as the new wagons emerged with six-wheel bogies. The flasks turned out to weigh around 50 tons, giving a gross laden weight of around 80 tons, within the permitted axle load for most of the network with just four axles. These pioneer flask carriers, also known as 'flatrols' and designated as FJ, were rebuilt with standard four-wheel bogies in 1981 and converted to air brakes, becoming XKB and then FNA under TOPS. The FNA design went through several upgrades, the most radical being in the early 1980s when it was decided that the flasks would be covered, giving an extra level of security, making it more difficult to access and impossible to tell if there actually was a flask on board. It also gave an extra level of containment should any radioactive material leak from the flask and, if it did, it would not be exacerbated by being washed by rain into the flatrol well or even on to the track.

The trains were shaped by the nuclear industry, literally! Magnox fuel elements were around a metre in length so were loaded into an almost-square metal box called a 'skip' and the flasks that carried them followed that shape, basically appearing as large distinctive white cubes. The second reactor phase, the AGRs used small ceramic uranium oxide pellets as fuel, encapsulated in bundles of long stainless-steel pins. However, as they were of a similar length to the Magnox fuel rods, a similar cuboid flask design was produced. The water-cooled reactors that sprang up around the world had many different forms, but their fuel assemblies were generally four or five times longer than their gas-cooled counterparts, spawning long cylindrical flask designs.

All flasks, however, have some basics in common, not least the ability to shield people from the lethal levels of ionising radiation, typically with 370 mm (15 in) of steel or various combinations of steel, lead and other materials to absorb different types of radiation. Most flasks also have cooling fins extending from the body, which help to dissipate heat generated by radioactive decay (the process by which an unstable atomic nucleus loses energy by radiation and heat generation). Most flasks are forged from a single steel casting, which gives tremendous strength, and hundreds of them were manufactured just down the road – or just down the Cumbrian coast railway line – from Sellafield at Workington. Most locals would describe the plant that made the flasks as Distington Engineering or just Chapel Bank, where it is located, although it lost that name in 1979 when it began to go through the full gamut of name changes with the rest of the steel industry. In 2017 it became TSP Engineering Ltd., a subsidiary of the British Steel consortium, Greybull.

In BR days flask trains had barrier wagons and brake vans. Brake vans were already disappearing from air-braked trains before privatisation and had disappeared from nuclear trains by 1998. Also, in BR days, nuclear flask trains had barrier wagons between the load and the locomotive, and between the load and the brake van, utilising whatever wagons were spare, typically long wheelbase vans and domestic coal hoppers whose original use had declined. Barrier or spacer wagons are also used to spread the weight of a train over bridges and viaducts dependent on the routes and loads carried. Barrier wagons are still used in certain circumstances but the criteria governing their use are not clear. According to the Network Rail (NR) document *Working Manual for Rail Staff Handling and Carriage of Dangerous Goods*, there should be six metres between a loaded nuclear flask and a locomotive. Since the latest FNA wagons are less than 12 m over the headstocks it seems impossible to meet that criterion. DRS probably adhere to a safe system of work agreed with NR but are reluctant to explain it.

A flask in transit on a mixed-goods train, the 0325 Carlisle–Stoke Gifford express freight passing through Wellington (Salop) on 28 September 1965. The wagon is a six-axle MJ flatrol, built at Swindon in 1960. (Courtesy Bill Wright)

FNA flatrol B900525, rebuilt at Shildon in 1981, with standard two-axle bogies and fitted with air brakes. This type of open flatrol characterised nuclear flask trains from the 1960s to the 1980s. Note the redundant sun canopy, which was fitted to loaded Magnox flasks to prevent additional solar heating of the flask. Spent Nuclear Fuel (SNF) still generates decay heat and must be kept cool; the fins on the flask effectively increase its surface area, assisting the heat to dissipate. After 1986 these open wagons were phased out in favour of (or converted to) covered types.

This is one of the flatrols that were converted and fitted with covers which slide clear to access the flask and a crude fillet has been welded into the gap below the cover. Locomotive 47363 heads the single flatrol and brake van over Arnside viaduct. As may be seen, in February 1985 the Kent Estuary was choked with ice.

By the mid-1980s new-build covered flatrols for use on domestic flask traffic were being introduced. This one – FNA 550025 – was built at Swindon in 1986 and is seen at Sellafield that year with its sliding cover partially open.

Pictured in 1985, this XJA flatrol was one of the first to incorporate a cover, which was lifted on and off by crane. The wagon was converted from the BAA fleet of steel carriers and carried a flask of fuel assemblies from the Steam Generating Heavy Water Reactor (SGHWR) at Winfrith.

Locomotive 40122 stands at Workington with a single FNA. In BR days flask trains had barrier wagons between the load and the locomotive, and between the load and brake van, using whatever wagons were available, long wheelbase vans in this case.

Locomotive 47522 passes Parton with a Carlisle–Sellafield flask train in September 1994. The barrier wagons in this case are coal hoppers, but towards the front of the train there are some with the hopper structures removed. This work was carried out by Currock wagon repair depot to produce the first bespoke barrier wagons.

Under sectorisation, nuclear was part of the coal sector. The evening combined Sellafield–Seaton and Fairlie flask trains have been allocated correct sector; locomotive 31200 passing Bransty signal box, Whitehaven, in September 1994.

Civil Link was the infrastructure sector in the late 1980s. Civil-liveried locomotives 31188 and 31255 head a flask train for Sellafield through Grange-over-Sands on 1 March 1996. Locomotives from other sectors appeared on flask trains, Civil Link in particular, because that sector optimised its motive power at weekends.

Regional Railways passenger sector locomotive 31455 was allocated to the Torness flask train on 29 February 1996, pictured leaving Sellafield.

The year 1994 saw BR create three freight companies, Loadhaul, Mainline and Transrail, supposedly to pave the way for privatisation, although they were all bought by what would become the English, Welsh & Scottish Railway (EWS) in 1996. Sellafield and the north-west of England fell into the Transrail area, but those regions disappeared when EWS took over. Here, Mainline-liveried locomotive 37077 leads 7A73, the 1721 Sellafield–Willesden over the viaduct at Eskmeals on 9 July 1997.

Service 7C48, 1721 Sellafield–Carlisle leaves Whitehaven on 19 May 1998. EWS is firmly established as the operator, with the company livery applied to locomotive 37667. By this time brake vans had disappeared.

Not all DRS trains had multiple locomotives. Locomotive 37611 heads service 7C48 at St Bees on 8 July 1999. EWS owned the bespoke barrier wagons, so it took them when they left! DRS tried various replacements such as the PXAs seen here.

A pair of DRS's second batch of locomotives, 20310 and 20314, heads service 7C22 the Carlisle–Sellafield flasks at Parton South junction on 9 June 1999. When DRS took over the home board flask trains, they were run in the same way that BR and EWS had done, combining two or more trains and tripping them via a hub such as Crewe or Carlisle.

Although it is difficult to find any evidence of an official policy, DRS ran its trains with two locomotives and often used the trains to reposition locomotives. Here, on 13 May 2015, locomotive 20308 is hitching a ride behind locomotives 37612 and 37611 south of St Bees with service 6C22, the 0656 Kingmoor–Sellafield.

DRS employs two locomotives on most of its trains. Where the route has a reversal (as at Morecambe on the way to Heysham) one locomotive is attached to each end of the train, a practice known as 'top and tail'. Here, on 1 June 2012, at Bare Lane on the Morecambe branch, locomotives 20304 and 20303 top and tail 2C41, the Sellafield–Heysham flasks.

Two locomotives provide insurance against a breakdown, a practical operational function in the privatised railway era where locomotives are few and far between. Here, locomotives 20310 and 20313 head 6E44, the 0745 Carlisle–Seaton flasks, past Brampton Fell signal box on 16 June 2008.

Locomotives 37087 and 37069 head service 6E44, the 0741 Carlisle–Seaton along the Durham Coast near Hawthorn Dene on 22 August 2011. Two locomotives make a breakdown less likely, thus avoiding the perceived security risk and public anxiety such an event may cause. However, in 2006 a flask train on this line triggered a hot axle box detector and came to a standstill in Sunderland station resulting in a nuclear scare according to local media.

With the partially frozen River Irt to the fore, locomotives 66420 and 66427 approach Drigg on 22 December 2010 with 6C53, the 0626 Crewe–Sellafield flasks. On the right of the picture Scafell rises above the rest, with Illgill Head, Lingmell and Great Gable in their winter coats.

The fourth version of FNA flask wagons, coded FNA-D, built by WH Davis at Shirebrook, are seen on a proving trial on 9 July 2014. Locomotives 57010 and 57002 top and tail service 6Z28, the 0703 Kingmoor–Derby Litchurch Lane, leaving the singled section of the Cumbrian Coast Line at Parton South Junction.

Locomotives 37606 and 37259 head 6C22, the DRS 0656 Kingmoor–Sellafield at St Bees on 27 February 2015. The weathered, in service FNA flasks contrast markedly with the rake of pristine FNA-Ds tagged on behind.

The flask transfer facility for the Hunterston B (AGR) power station is sited in the Hunterston Ore Terminal on the Firth of Clyde. Locomotives 68022 and 68029 are seen soon after arrival in the Hunterston Causeway Siding where the flask unloading gantry is located. The train is 6S54 the 0442 from Kingmoor. (Courtesy Adrian Nicholls)

Locomotives 88001 and 88002 power 6C53, the 0625 Crewe–Sellafield flask train at Green Road on 20 February 2018. Although still distinguishable by their green bodysides the FNA-Ds are becoming as weathered as the FNA-Cs.

Chapter 3

Transport of Spent Nuclear Fuel: Imported

As mentioned earlier in Chapter 1, the reprocessing of SNF was a significant source of foreign currency and became the nation's biggest earner of yen. Japan had imported its first commercial reactor from the UK, a GEC Magnox design that operated successfully at Tokai Mura from 1966 to 1998. Japan subsequently bought American-designed light water reactors (LWRs) and then developed its own versions; however, it continued to send SNF to Sellafield using Barrow-in-Furness as the main port of access. As the LWR fuel assemblies were much longer than the home board fuel elements – up to five metres in length – a longer flask was needed, and a design was conceived to accommodate them. There are various designs of LWRs, the two most common types are the Boiling Water Reactor (BWR), where the water in the reactor vessel turns directly to steam, and the Pressurised Water Reactor (PWR), which keeps the water at high pressure allowing it to get hotter without boiling.

The main group of flasks built to transport the LWR fuel was led by the French-designed Excellox. The fuel elements were placed in a long cylindrical stainless-steel tube, called a Multi Element Bottle (MEB). The MEB fitted inside the flask, which itself was basically a long cylinder, and both were filled with water to assist with conducting the decay heat away and provide neutron shielding. There were several designs similar to the Excellox, among them the Nuclear Transport Ltd. (NTL) versions, which tended to be used for European transfers. All these flasks had thinner steel shells than the original cuboid Magnox flasks to allow more lead to be inserted, as LWR fuel tended to stay in the reactor longer so became more radioactive.

NTL was an international European consortium comprising interests from the UK, France and Germany to cooperate in the movement of (mainly) LWR fuel between European reactors and Europe's two reprocessing facilities at Sellafield and Cap De La Hague in northern France. Some of these flasks were also made at Distington Engineering works at Chapel Bank, Workington. French manufacturer Fauvet Girel built the ferry wagons to carry the NTL flasks in the late 1970s, an eight-axle, double-bogie design complete with sliding flask covers. These wagons plied the North Sea and channel ferry routes until the mid-1990s, when the Channel Tunnel put the rail ferries out of business. Larger British wagons for SNF that was imported through Barrow docks were also eight-axle double-bogie designs but were open and had interchangeable frames able to carry two 51-ton Magnox cuboid flasks or a single 100-ton cylindrical flask mounted horizontally.

Pacific Nuclear Transport Ltd. (PNTL) is a similar consortium to NTL, concerned with transport between Japan and the two European reprocessing facilities. It operated three nuclear transport ships, *Pacific Heron*, *Pacific Grebe* and *Pacific Egret*. Since the inception of the NDA yet another authority, International Nuclear Services Ltd. (INS), has been tasked with overseeing all international nuclear transport on behalf of the UK government. INS holds a stake in PNTL and owns the nuclear cargo ship, *Oceanic Pintail*.

The UK did dabble with water-cooled reactors, building the Steam Generating Heavy Water Reactor (SGHWR) at Winfrith in Dorset. The reactor was moderated using heavy water, and cooled by light water, operating like a BWR. One of the problems with the use of light water as a moderator is it absorbs neutrons. The nucleus of hydrogen has a proton but no neutron, so tends to absorb one.

The hydrogen in heavy water already has the neutron so 'ignores' the passing neutrons, which are needed to maintain the chain reaction in the reactor core. Although it was an experimental reactor the SGHWR did supply power to the National Grid from 1968 until 1990.

DRS took over responsibility for imported irradiated nuclear fuel in September 1996, even before they started managing the home board traffic.

Another flask wagon with removable covers is seen here in 1983, passing through Workington behind locomotive 25244. The flask was from an experimental Boiling Water Reactor (BWR) at Dodewaard in The Netherlands, entering the country through Humberside (variously between the port of Hull and Immingham).

Although small, the Dodewaard flask still required heavy shielding, causing a distinct bowing in the centre of this XNB wagon.

Built at Ashford in 1977, and originally designated PIA, this is a PXA double-bogie wagon loaded with two Magnox flasks, seen outside the Sellafield plant on 22 July 1996. The SNF in these flasks was imported through Barrow Docks from the Tokai Mura power station, Japan. Each flask weighs 51 tons, but with the frames and shock absorbers the pair form a 105-ton payload. The total gross laden weight is a little over 165 tons.

This is the same type of eight-axle wagon as in the previous picture, but designated KXA, and loaded with a single cylindrical Excellox flask mounted horizontally. The V-shaped frames above the outer bogies on the wagon are for internal use on the Sellafield site to transport the circular metallic grey shock absorbers, seen here mounted on the flask ends.

A trainload of SNF from Japanese PWRs, imported through Barrow docks and bound for Sellafield. After running around, locomotive 25325 has propelled the train out of the docks branch and is getting underway from Salthouse Junction in February 1985. The vans in the trainload are to spread the weight of the whole train on the viaducts.

In 1987, BR divided its freight operations to concentrate on trainload working and dedicate resources to the commodities carried. The nuclear operations became part of the Trainload Coal sector, represented by the black diamonds logo seen on the bodyside of locomotive 31276. The train, pictured at Barrow Ramsden Dock, is made up of empty flasks returning to Japan.

DRS took over responsibility for imported SNF in September 1996, even before it started managing the home board traffic. Locomotives 20302 and 20305 (from DRS's first batch of locomotives) head a Sellafield–Barrow Ramsden Dock train through Foxfield on 26 November 1996.

In July 1997 DRS purchased six Class 37 locomotives from the failed Nightstar project and pressed them into service immediately, still carrying Eurostar livery with Eurotunnel cast roundels. Locomotives 37609 and 37610 are passing Ormsgill, Barrow, in July 1997, with two KXA wagons and a flask from Switzerland.

One of the French-built nuclear flask ferry-wagons that were designated IQE or IQA and which transported SNF from the continent by rail ferry. This empty one is marshalled into an ordinary freight train, the 1234 Workington–Warrington Arpley, powered by locomotive 47308, leaving Sellafield in July 1995.

An eight-axle ferry-wagon, 33879985002-3, built by French manufacturer Fauvet Girel. The wagon was in Workington Yard in 1985, possibly being used to transport a new flask to BNFL, and which had been manufactured by the British Steel-operated Cumbria Engineering, referred to locally as Distington Engineering or Chapel Bank.

Chapter 4

Escort Coaches

Most of the nuclear material carried by rail is hazardous but cannot be used to make nuclear weapons, and the flasks are so difficult to penetrate it is difficult to see how criminals or terrorists could realistically make use of it. However, some materials moved on the railways do need to be escorted for security reasons.

The railways have played virtually no part in the weapons programme but do have a role with regards to nuclear submarine propulsion systems. Great Britain's nuclear missiles (and the vessels that exist to deliver them) are based mainly on American technology. It is no secret that nuclear submarines generally are powered by PWRs, and fuels such as highly enriched uranium are needed to make them smaller and able to operate for extended periods. The SNF from the submarines is occasionally sent to Sellafield by train in flasks accompanied by escort coaches.

The Civil Nuclear Constabulary (CNC) – according to its website – is an armed police force that is responsible for protecting nuclear material. The initial position of this police force is to deter any attacker whose intent is the theft or sabotage of the material in transit but, if attacked, will defend it. As might be expected on-board CNC police officers are not the only thing would-be attackers would have to contend with; when these trains run there is also considerable additional support away from the trains.

There are a few other materials that have needed escort coaches, including material being moved from Dounreay to Sellafield. The Dounreay site is being decommissioned and –Sellafield apart – is probably the most diverse and complex nuclear site in the UK. Owing to the experimental and diverse characteristics of some of the materials at Dounreay they are often referred to as 'exotics'.

Excluding some notable one-offs over short stretches of line, the largest wagon to run on Britain's railways over any distance was a flask carrier designed to take SNF from the naval yards at Chatham, Rosyth or Devonport to Sellafield. As submarine reactor cores became smaller it became easier to design a smaller flask and wagon. Although smaller, these are still massive vehicles: the latest incarnation (designated KUA) is derived from the covered ferry-wagon concept with eight axles.

The PXX was the largest wagon to run over long distances on Britain's railways. A Class 25-powered Sellafield–Devonport flask train leaving Sellafield in April 1985; the escort coaches were designated PPX though later had TOPS code KCX.

Numbered MODA95780 this PXX was built by Head-Wrightson in 1963, to a basic transformer transporter design. The flask was suspended vertically and, although around 15 cm from the railhead, looked like it was almost touching. More than 27 m in length (89 ft 5 in) it weighed 94.4 tons and with the 110-ton flask its gross laden weight was over 204 tons, supported by the twelve axles of its double bogies. For much of its journey it would travel as an out-of-gauge load.

At Sellafield on 24 July 1996 is locomotive 47293 with two flask wagons from the Rosyth Naval dockyard with escort coaches. The old Mk 1 escort coaches and the massive flask transporter have given way to slightly smaller versions. The flask carriers were based on the French-built internationally registered ferry-wagons. The escort coaches are short-framed Mk 1 inspection saloons; the one behind the locomotive is an old London Midland & Scottish (LMS) vehicle from the 1940s. The wagons are 33879985008/9, the coaches DB999509 and DM45020.

In 1998, the IQA derivatives were replaced by the KUA, built by Bombardier Prorail of Wakefield. This one, alongside the River Ehen at Sellafield on 9 July 2011, is in 6C46 the 1237 Sellafield–Kingmoor service, along with two standard FNAs and led by locomotives 37229 and 37409.

KUA flask carrier MODA95770. It differs from the ferry-wagon derivatives most obviously with the covers that drop down further, sliding along a low rail just above the bogies.

When DRS took over the escorted materials traffic, it acquired some BR Mk 2 standard coaches. This one – seen with locomotive 20301 at Sellafield – is Mk 2 brake 35512 still carrying its Regional Railways livery and logos. Locomotive 20301 was regarded as DRS's first locomotive and is carrying a small commemorative nameplate, *Furness Railway 150*. This was removed in 2000 and the locomotive was re-named *Max Joule 1958–1999*, in honour of the company's first Managing Director, who was tragically killed in a cycling accident.

DRS eventually adopted and modified four Mk 2 coaches of their own. Coaches 9419 and 9428 are seen at Parton on 31 July 2014, sandwiched between locomotives 57009 and 57008. They are forming 5Z50, the 0913 Sellafield–Longtown ecs (off the 1847 Keyham–Sellafield flask train).

Locomotives 68017 and 68020 combined to power 6C72, the 0912 Barrow Marine Terminal–Sellafield towards Green Road on 25 April 2017 (locomotive 68001 is on the tail). The white containers are carrying material from Dounreay.

A dual refrigerated 20-ft container in 6C72, the 0914 Barrow Marine Terminal–Sellafield, at Kirkby-in-Furness on 29 May 2018. The container is mounted on an IKA semi-permanently coupled pair of low deck units, similar to those on intermodal service with DRS. Also shown in the picture is the Mk 2e escort coach 9508 and locomotives 68005 and 68002.

Chapter 5

Nuclear Waste

Nuclear waste is probably the most misunderstood of all the types of nuclear traffic on the railway system. SNF may be regarded as high-level waste, but this chapter looks at the waste separated after reprocessing, any unwanted material activated in a reactor or contaminated material produced by the nuclear industry, general industry, the Ministry of Defence (MoD), universities and medical facilities. There are four classifications of nuclear waste:

Classifications of Nuclear Waste

Type of waste	Definition
Very Low Level Waste (VLLW)	Material not considered harmful to people or the environment
Low Level Waste (LLW)	Radioactivity not exceeding 4 GBq/t α or 12 GBq/t $\beta-\gamma$
Intermediate Level Waste (ILW)	Radioactivity exceeding LLW levels but heat generation not exceeding 2 kW/m^3
High Level Waste (HLW)	Heat generation exceeds 2 kW/m^3 so requires cooling and shielding

Very Low Level Waste (VLLW) may be ignored as it is not routinely carried on trains.

Low Level Waste (LLW) needs little worry about the numbers or the units; the principle is, if the radioactivity is less than a specified amount per unit of weight, it may be sent to the LLWR at Drigg. Most of this waste is of lower hazard and requires no special shielding, although some precautions, like separating it from people and monitoring the packages, are required.

High Level Waste (HLW) is not categorised by how radioactive it is but by the heat it generates. Previous chapters have explained that radioactive decay produces heat so, if the waste generates heat at a rate that requires engineering measures to cool it and keep it safe, this must be classified as HLW.

The reprocessing of irradiated fuel gives rise to highly active liquors. At Sellafield these are vitrified in stainless-steel canisters and returned to their country of origin via Barrow Docks. This sort of waste needs heavy shielding as well as cooling. Higher Activity Waste (HAW) is mainly HLW, but it is a slightly broader category that includes lower category wastes such as ILW if, for example, it has properties that pose handling problems.

Intermediate Level Waste (ILW) does not generate enough heat to require additional measures like HLW but has too much radioactive material per unit of weight to be LLW. However, there are different types of ILW. Much of it requires shielding but some does not as long as it is contained, for example plutonium-contaminated waste. This is sometimes called Contact Handleable Intermediate Level Waste (CHILW).

Until 1978, LLW was transported by Dempster Dumpster-type road vehicles to the disposal facility at Drigg, where local residents were becoming increasingly vocal about the dozens of skips passing through the villages of Holmrook and Drigg every day. In 1977 the inquiry into the proposal to build THORP was in progress, during which BNFL gave an undertaking to transfer the skip movements to rail.

Between 1959 and 1995, 800,000 m3 of the waste from these skips was tumble-tipped into seven clay-lined trenches at Drigg. In 1988, construction of Vault 8 began. Instead of tipping waste directly into trenches the waste would remain in the transport container to be stacked in a concrete-lined facility. Around the time that construction of Vault 8 began, BNFL also began designing a Waste Monitoring and Compaction Plant, which received the acronym 'WAMAC'. So, not only were the containers now stored in a proper facility, but the waste was compacted to save space and the contents grouted to stabilise the package. In addition to compaction, new techniques in the nuclear industry, such as super compaction, thermal treatment and various treatments of metals, have reduced the amount of waste destined for the LLWR considerably; the days of daily waste trains are long gone. The LLWR at Drigg is the national repository and takes waste from nuclear facilities, general industry, the MoD, hospitals and universities. It is managed by UK Nuclear Waste Management Ltd., under the auspices of the NDA.

Many railway enthusiasts and commentators are unaware of the distinction between nuclear wastes but ILW and LLW are very different. ILW presents significant hazards to people and the environment so must be contained in robust packaging, and some types of ILW require significant radiation shielding. For these reasons ILW is managed and stored in purpose-built facilities at Sellafield and transported in specialised containers built to standard ISO dimensions (2.4 × 2.6 × 6 m), colloquially known as '20-ft boxes'. These boxes are used to transport ILW originating from the clean-up of old UKAEA experimental and research facilities, and reactor sites in the Gloucestershire, Oxfordshire and Dorset areas (Aldermaston, Culham, Winfrith and Harwell); the latter is the NDA centralised waste store for the area, using Berkeley as the railhead. The other container types for ILW were the smaller NUPAK packages, which were designed to carry four 200 litre galvanised steel or stainless-steel drums. Generally two of these are mounted on a PFA flat wagon via a transition frame.

Some ILW had been stored safely at Drigg for many years, but the containers holding it and the buildings where it was housed were not designed for indefinite storage, so since 2010 the material – mostly plutonium-contaminated waste – was recovered and returned to Sellafield where it was repackaged and securely stored.

An up-rated version of the NUPAK, called a NOVAPAK was introduced at the end of 2017 and the contract to build these was awarded to Bendall's Engineering of Carlisle. According to their website, each container has two layers of thermal shielding and impact protection and was fabricated from 1,500 components. Their construction required 750 welds, for which twelve welders were employed!

In Chapter 1, reference was made to the splitting of uranium nuclei into new elements, called fission products, that inhibit the flow of neutrons in the reactor – eventually forcing the fuel's replacement. Intensely radioactive, they emerge from the reprocessing procedure as a liquid, often referred to as Highly Active Liquor (HAL). At Sellafield, the vitrification process chemically bonds the radioactive content into a glass matrix, forming a stable solid that is impossible to leak, easier to store and much easier to transport.

By agreement, HAW is returned to the country that originally sent its SNF to Sellafield under the Vitrified Residue Returns (VRR) programme, going back the way it came, by sea through Barrow Marine Terminal.

In 1985, BNFL secured a contract to dispose of trace active liquors from the USA. The chemicals, which contained traces of radioactivity, were fed into the Sellafield reprocessing system instead of fresh chemicals. Here, on 16 August, a couple of containers approach Park South Junction behind locomotive 47398.

LLW at Sellafield was mainly protective clothing, tools and equipment contaminated in day to day operations. It was bagged and consigned to skips like the one on the right of this picture. Heavier items, such as metal, rubble, etc., were placed in heavy-duty versions like the skip on the left. The flat wagons used were designated PFA but were a more basic design and of longer wheelbase than the later version of PFA used by DRS.

Here, on 6 May 1993, locomotive 40194 is setting off with the daily LLW train (Target 60), which left Sellafield at the end of every working day.

A more robust type of reuseable container was tried in the 1990s but was short-lived because the industry was moving towards storage of LLW in single-use ISO freight containers. Containers like this, designated TC-05, remained in use for many years but saw little use on trains after 1995.

By the mid-1990s, LLW had been revolutionised; it was compacted at Sellafield and consigned to Drigg in ISO containers that were stored in a vault, in a much more stable and retrievable condition. Locomotives 20302 and 20305 head into the morning sun on 26 November 1996.

The Sellafield–Drigg LLWR trip utilised whatever locomotives were available at Sellafield. Because of the short journey (less than four miles) there was no requirement for speed. Here, Harry Needle's 08375 passes Seascale Golf Course on 29 April 1999, being trialled on the short trip.

In 1998, DRS bought a batch of Class 20s from Hunslet Barclay. Here, locomotive 20902 gets a 'nose-first' outing, on 9 June 1999, with 7C20, the 0807 Sellafield–Drigg, LLW train with two containers on a KFA-type bogie wagon.

The standard LLW train is made up of the DRS ubiquitous PFA flat wagons carrying the basic TC01 half-height (1.3 m high) container. Although they conform to the standard ISO profile, the TC01 boxes are a specialised item for permanent storage. They are manufactured by W H Davis, who provide several types of ISO container including TC03, a one-third-height heavy duty variant for rubble or soil. The white exhaust emanating from locomotive 37425 indicates that it has not been warmed up for long on this frosty 28 February 2013.

The TC02 half-height container is a reusable design to transport material that has already been properly pre-packaged, or for very large items. One is seen here on the rear of the Sellafield–Drigg train at Seascale, on 15 March 2011. It carries no labels, indicating that it is empty and will probably be used to remove something from Drigg. The train is mainly made up of standard TC01 boxes with some green NUPAKs behind locomotive 37229.

Occasionally, some large items of LLW are carried in full-height containers like this one being drawn into Workington Docks by locomotive 57010 on 30 January 2013. They carry items such as contaminated metal from the LLWR at Drigg, which may be decontaminated by specialised processes leaving the metal to be recycled normally. The boxes will be transported by road to a specialist processor in Workington. At that time (2013) this was a Swedish company, Studsvik, who were taken over by EDF (Électricité de France).

Typically, a full-height ISO freight container will be used for ILW. From 2015 these special containers started to be seen regularly on the Cumbrian Coast Line, as here, where locomotives 68026 and 68023 combine to power the 0130 Crewe–Sellafield service past Nethertown on 19 July 2017. The full-height container – on its frame – is deemed to be out of gauge for the tunnels on the Furness line, so is routed via Shap and Workington.

Frequently confused with LLW, this is a consignment of ILW on 6C53 near Harrington, in August 2013. The picture illustrates the use of transition frames to carry containers on PFA flat wagons. The container in the centre of this picture is mounted on a relatively slim metal frame, which can be seen between the wagon and the container. (Note there is no frame on the empty wagon to the left.) The PFA on the right carries one of the smaller NUPAK containers, which needs two frames: the same thin frame as the ISO and the larger (light blue) frame to convert the standard ISO fitting to the non-standard darker blue box.

This picture was taken on 21 March 2018, at the Dungeness railhead in Kent. Top and tailed by locomotives 37059 and 37069, it shows a typical train of full-height ISO containers of the type used to transport ILW between stations in the south of England. Rather than have each station store a full range of different wastes, the NDA planned to create specialist facilities on existing nuclear sites to consolidate similar wastes. Because of the perceived risks in transporting the wastes there was some public opposition to the plan. (Courtesy John Waddington)

Locomotives 68034 and 68004 top and tail 6Z68, the 1130 Kingmoor–Hedon Road Sidings on 17 October 2017. This is a return of UO3 through Hull docks to Russia after reprocessing. The train is seen on the Newcastle–Carlisle line, east of Scotby, on the outskirts of Carlisle.

This is *not* LLW going from Sellafield to Drigg, the empty NUPAK containers will carry ILW the other way. Starting in 2010, material began to be recovered and returned to Sellafield where it was repackaged and securely stored. This is locomotive 37682 ambling past Seascale Golf Course in July 2011, heading 7C20, the Sellafield–Drigg trip.

In 2017, the NOVAPAK, built by Bendall's Engineering of Carlisle, was introduced to replace the NUPAK. Here, a pair of NOVAPAKs are in the consist of 6K74 at St Bees on a rainy 10 September 2018.

Marshalled behind locomotives 88010 and 88002, in the consist of 6K74 at St Bees on 7 August 2019, is PFA 92736 carrying another type of container associated with material from the decommissioning of the Winfrith Dragon Reactor.

Service 7X23, the 0932 Sellafield–Barrow Marine Terminal, passes Sowerby Lodge on the single line section from Park South junction to Barrow-in-Furness on 13 February 2014. The train, comprising two KXA eight-axle flatrols each loaded with a TN28VT flask, is top and tailed by locomotives 37608 and 37405 carrying HAW.

TN28VT flasks arrive at the Barrow Marine (formerly Ramsden Dock) Terminal on 13 February 2014 in the care of locomotives 37608 and 37405. They will be loaded on to *Pacific Grebe* for their journey to Japan, returning HAW under the Vitrified Residue Returns (VRR) programme.

On 16 October 2016, locomotive 37609 has been detached from the head of the train (6X23, the 0625 Sellafield–Barrow Marine Terminal) and run into the spur to allow the tail locomotive (37218) to propel the flask train into their destination. The flask is a Castor HAW 28M, returning HAW to Switzerland aboard *Oceanic Pintail* via Cherbourg.

This picture offers a good opportunity to marvel at the beautiful engineering on display; in particular, the graceful curves of the W H Davis-built KXA-C wagon, both longitudinally and in its cross sections. One can see that the wagon has a narrow bed (broadened only in the centre to support the flask) allowing it to meet the network loading gauge despite its length. It has a tare weight of 53.5 tons and can carry a 126.5-ton payload (180 tons gross), and its eight axles maintain the 22.5-ton/axle loading. Mounted on the wagon is a Castor HAW 28M-type flask weighing around 112 tons.

Locomotive 37218 shunts 6X23 past the neatly manicured lawns outside the Barrow Marine Terminal on 6 October 2016. The terminal is operated by International Nuclear Services (INS).

Locomotive 57305 shunts Barrow Marine Terminal on 20 June 2018, not long after DRS retired these Pullman-liveried Class 57s from Northern Belle luxury excursion duties. The Castor HAW 28M flasks are returning empty to Sellafield from Germany.

Chapter 6

Chemicals

Sellafield is a large nuclear plant, but it is also a complex chemical plant too, consuming large quantities of chemicals. Its main rail-borne chemicals were nitric acid and sodium hydroxide. Sellafield's chemicals were pumped around the site from a tank farm located next to internal rail sidings. Sodium hydroxide, delivered in the form of caustic soda liquor, had many uses at Sellafield but was utilised primarily to manage pH levels (alkalinity) throughout the site's storage ponds and chemical plant. Nitric acid also had many roles at Sellafield but was used mainly in the separation process of dissolving nuclear fuel.

Both Sellafield and BNFL's fuel manufacturing facility at Springfields used BR wagonload freight trains to deliver their chemicals. In 1987 BR divided its freight operations to concentrate on trainload working and dedicate resources to the commodities carried. The nuclear operations became part of the Trainload Coal sector and chemicals became part of the Trainload Petroleum sector. The next upheaval came in 1994 with the creation of three regionally based freight companies, Loadhaul, Mainline and Transrail, which were supposed to pave the way for privatisation. Sellafield and the north-west of England fell into the Transrail sphere. No sooner had the three freight companies started painting their stock in house colours when, in 1996, all three companies were bought by a consortium led by Wisconsin Central, eventually becoming English Welsh & Scottish Railway (EWS). The head of this operation tried to re-establish wagonload operations and the Sellafield chemicals went back to deliveries based on general freight trains while the nuclear trains continued as before, the only difference being the plethora of different liveries to be seen.

Chemicals was the first revenue-earning commodity that DRS took over from EWS in 1996, although the relationship did not end there as EWS was still taking chemicals to Carlisle Yard for DRS to pick up and set down well into the twenty-first century. Procurement of chemicals for both Springfields and Sellafield was centred on many sites in the Cheshire area: the Kemira site at Ince near Helsby; the Hays site near Sandbach, and the ICI–Chlor site at Folly Lane near Runcorn. In the summer of 2011, Sellafield Ltd. began to source nitric acid from Terra Industries at Wilton on Teesside, which continued into 2012 but was seen less regularly into 2013 when Sellafield Ltd. brought chemical deliveries by rail to an end.

Sellafield is a complex nuclear plant but also a very large chemical plant. In BR days Sellafield was served by various wagon load or trip freights to deliver chemicals. Here (in April 1995) are some typical 51-ton gross laden weight TTA caustic soda tanks in the Workington–Warrington Arpley freight, being shunted by locomotive 47193 in Sellafield's up-sidings.

Nitric acid had many roles at Sellafield but was used primarily in the separation process of dissolving SNF. Locomotive 37108 hurries the 1028 Sellafield–Ellesmere Port East Sidings Kemira tanks through Kents Bank on 24 July 1993.

Chemicals was the first revenue-earning commodity that DRS took over in 1996. Here, at Ravenglass, in February 1996, locomotives 20301 and 20302 head the train of nitric acid from Ince in Cheshire.

6C43, the 0925 Sellafield–Carlisle Yard caustic soda train, consisting of only one tank, rounds Redness Point, Whitehaven, on 14 June 2003. In typical DRS traction-to-train imbalance, locomotives 20305 and 20304 head the train with locomotive 37059 inside.

This is 6C43, the 0925 Sellafield–Carlisle Yard caustic soda tanks at Parton in September 2003. Locomotive 33025 carries 'minimodal' branding, a concept discussed in Chapter 8 of this book.

Locomotive 37602 heads 6M24, the 1642 Middlesbrough–Kingmoor tanks by the River Tyne at Wylam, on 28 August 2012. Within a year, chemical trains serving the nuclear industry would end.

Chapter 7

Infrastructure and Construction

When Vault 9, the new nuclear waste facility at the LLWR, opened in 2010 the NDA website posted that the decision to move the vast majority of construction materials by rail rather than road eliminated a staggering 27,500 potential road deliveries through the village of Drigg (and nearby Holmrook). The port at Workington acted as a railhead for local quarries to stockpile large quantities of aggregates and also some general building materials from local suppliers. Stone was also sourced from Ghyll Scaur quarry near Millom, where trains were loaded whilst standing on the main line under a night-time possession. DRS was also contracted to transport aggregates to the LLWR at Drigg in 2015–2016. A stockpile was created on Associated British Ports land at Barrow Docks by Burlington Aggregates Ltd. (an amalgamation of local firms Neil Price Construction Services and the aggregates division of Burlington Slate Ltd.).

BR had run some similar trains of construction aggregates from Workington Docks to the BNFL Drigg site in 1992, but the biggest project supported was the THORP project, at the time Europe's largest construction programme. Although work started on the site in 1981, it was not until 1984 that the major civil work was underway. Thousands of tons of reinforced concrete were needed for the construction of the THORP plant at Sellafield, and this was mostly delivered by the Warrington–Workington 'pick up' freights with additional trips between Sellafield and the freight hub of Workington. A small, temporary cement terminal with a horizontal buffer silo was set up in the head-shunt of the upsidings at Sellafield, where cement was pneumatically transferred to trucks for movement to the site. Similarly, a mobile crane was deployed at the sidings loading the steel reinforcing bars (re-bar) to flatbed road vehicles.

DRS has also secured contracts for infrastructure support on the railway with Network Rail. This support includes ballast trains, measurement trains and in the autumn months railhead treatment trains.

A small, temporary cement terminal was set up in the head-shunt of the up-sidings at Sellafield during the construction of THORP, seen here in 1986.

Locomotive 25201 heads 7P36, a typical pick up freight near St Bees in the early 1980s, including cement tanks related to the THORP construction project at Sellafield.

Steel reinforcing bar (or 're-bar') for THORP's construction was delivered in large quantities; in 1986 a Class 31 locomotive is swapping loaded BDAs for empties.

Here, re-bar is being loaded on a truck for the short road transfer into the Sellafield construction site.

Below: The nuclear industry continued to use railways to deliver construction materials after privatisation. The construction and capping of nuclear waste storage facilities, known as 'vaults', at the LLWR generated many aggregates trains like this one at Moss Bay, Workington on 8 December 2009. Locomotive 66414 is in a Stobart Rail livery associated with intermodal operations (covered in Chapter 8 of this book).

Here, at Kirkby-in-Furness on 21 February 2018, DRS provided locomotive 66425 to power a train of empty JNA bogie box wagons (hired from GBRf) to transport quarry waste from a stockpile at Barrow Docks to be used to cap trenches and vaults at the LLWR. The train is 6C21, the 1125 Sellafield–Barrow Docks with locomotive 66432 on the rear.

Not only aggregates were carried to the construction site at Drigg; here, some other building materials leave Workington Docks in November 2008, top and tailed by locomotives 37038 and 37609.

It is not only the nuclear industry that DRS supports with infrastructure trains. Here, a contract from Network Rail for ballast sees locomotives 37611 and 20302 top Shap summit on 17 December 2013 with 6C27, the 0942 Carlisle VQ–Shap Quarry.

DRS also participates in the scheduled network of departmental trains run on behalf of Network Rail. This one is 6K05 the 1246 Carlisle Yard–Crewe Basford Hall (on this occasion a High Output Ballast Cleaning unit) powered by locomotive 68034 in DRS livery and Scotrail-liveried locomotive 68006. The train is crossing Ais Gill viaduct on 28 September 2018.

Here is a ballast train in action, utilising the ballast from the virtual quarry at Carlisle Yard brought to the site by locomotive 57004. Following storms and tidal surge on the Cumbrian coast at Siddick, on 9 January 2014, the formation had been washed away.

During the leaf fall season DRS provides traction for rail-head treatment trains all across the network. Having just passed Milton Level Crossing on 16 October 2017, locomotive 37602 heads west on the Newcastle–Carlisle Line with 3S77, the 0515 Kingmoor–Kingmoor Water cannon set, which treats the North East area (locomotive 37609 was at the rear).

On 8 October 2019, locomotives 66301 and 66302 top and tail 3J11, the 1621 Kingmoor–Barrow–Kingmoor RHTT, seen between Parton South and Parton North junctions, on the singled section of the Cumbrian Coast Line below Lowca Sea Brows.

Other Network Rail contracts supported by DRS include measurement trains like this one. Locomotives 37402 and 37409 top and tail 1Q82, NR's 1537 Carlisle–Blackpool North, crossing the Derwent viaduct, Workington, on 28 March 2019.

Chapter 8

Intermodal

Intermodal is a success story for DRS. However, this is probably the right time to pause and ask the question: why do DRS get involved in business not connected to nuclear traffic? Well, the answer lies in its corporate strategy, which is accessible via their website. The objective of DRS is stated as: "Security of supply for nuclear transport is our primary focus", which is a concise, clear goal. The rest of the strategy document is not as sharp as that, and combines a lot of detail with overly long sentences, paraphrased here: 'No-one knows how the nuclear industry will change going forward so, if DRS is to meet its objective, it must maintain knowledge and expertise about the national rail system. Similarly, the company must keep abreast of the developing technologies in locomotives, rolling stock, railway infrastructure, railway business practice and logistics.'

This means running trains outside the limitations of only running nuclear traffic on limited routes. However, DRS cannot run any trains it fancies, the strategy document also insists that the taxpayer must get value for money and that new contracts meet their existing commercial, safety, security and environmental standards. Finally, DRS must recruit and keep the best people; the growth and diversification of the business, as outlined above, is part of the strategy to help to do just that. In short, DRS must have quality people, operate railway facilities across the network and be able to run just about any sort of train, anywhere, at any time.

So, back to intermodal; DRS had their first shot at intermodal in June and July of 1997. It partnered with a dairy farmers' co-operative called Milk Marque in a trial that used 'piggy-back' wagons to transport milk from Cumbria to the capital. A specially designed Exel–Tankfreight road tanker was loaded with 28,000–30,000 litres of milk every afternoon and carried on a piggy-back wagon from Penrith every day for four weeks to Cricklewood behind a pair of Class 20 locomotives (or, on occasion, newly acquired Class 37s). The road tanker was unloaded at Cricklewood and driven to a dairy, where its contents were discharged. The tanker returned to the depot to re-join the train, which returned overnight to Penrith. The trial ran under the title of 'MILKLINER 2000' and, from an operational perspective, was a complete success. This project was being developed into a full-blown operation when it was curtailed due to an unfortunate intervention by the then Monopolies and Mergers Commission. For reasons unconnected to the piggy-back trials the Commission insisted on the break-up of Milk Marque and, as a consequence, the 'Milkliner' project went down with it.

DRS has supported several intermodal initiatives. In August 2002, some demonstration trials were conducted with another concept, dubbed 'mini-modal', in which a mega-fret type of intermodal wagon was used to carry six small cube-shaped 2.5 m containers. The boxes had roller shutter doors so they could be loaded or unloaded at platforms or – for greater security – be placed on the wagon with doors facing each other. Individual boxes could be unloaded at yards or even rural stations using equipment such as agricultural telehandlers, which may be accessed in remote rural areas such as stations on the Settle–Carlisle or Highlands lines.

In 2011, DRS was involved with Britain's 'Energy Coast' (a consortium of NMP, local enterprises and national and local governments) in promoting an upgrade of the port of Workington to an international intermodal hub. The initiative was called MULTIMODAL CUMBRIA; its intention was to transform Workington into the only container port between the rivers Mersey and Clyde and, to that end, a giant Liebherr crane was purchased. Unfortunately, the initiative did not bear fruit (at the time of writing in 2020). However, the possibility of such business remains open.

DRS began operating what might be described as a normal intermodal service in 2001 for WH Malcolm, between Grangemouth and the Daventry International Rail Freight Terminal (DIRFT) in Northamptonshire. Trials over this route had been taking place over the preceding year and even earlier with the Stobart Group. Back in 1998 the Stobart trials were held in complete secrecy: the reason given for the trains was for DRS to evaluate multiple working between classes 20 and 37 locomotives. This subterfuge worked quite well until the Stobart Group decided to paint one of the wagons in its distinctive Eddie Stobart livery! By 2002, DRS had received its first batch of Class 66 locomotives and the trains were proving reliable; things were looking up.

Looking back at those partnerships from many years later it seems unexceptional, but at the turn of the century it was ground-breaking for a rail company to work so closely with what had traditionally been thought of as a road haulier. The Malcolm Group, the Russel Group and the Stobart Group had become logistics businesses providing door-to-door services in co-operation with DRS (and indeed other rail companies). They served a range of customers, in particular supermarkets Asda, Morrisons and Tesco. DRS formed a partnership with fellow Carlisle-based logistics firm the Stobart Group and turned out locomotives 66411 and 66414 in a Stobart Rail livery. Locomotive 66412 received a Malcolm livery in 2008 and, after that locomotive went off lease, it was replaced by locomotive 66434, which received a different version of the Malcolm livery in 2012.

The trains proved to be very reliable and were well loaded in both directions. A lot is made of the environmental benefit of rail over road with the clever use of the 'LESS CO2' motif resembling the 'TESCO' logotype. Clearly, less CO2 *was* generated as the trains took 130,000 lorry journeys off the roads each year, but DRS received some criticism for using diesel locomotives over an electrified route. In 2010, DRS lost control of the Tesco train to rival company DB Shenker for a short period. DB Shenker ran the train with its Class 92 electric locomotives, which (to be fair) was more in keeping with the environmental credentials claimed for this train. When DRS won back the contract it had the audacity to hire the same locomotives from DB Shenker! DRS preferred the flexibility of diesel locomotives though, as it allowed the use of diversionary routes for both planned and unplanned line closures, as getting the goods through was a crucial element of its ethos. Eventually, DRS went back to Class 66 locomotives but, in 2015, started to trial Class 68 diesels on the Tesco train temporarily, until the Class 88 bi-mode locomotives became available in 2017.

DRS clearly thought the alternative route issue through carefully: the low 2.59 m high (8 ft 6 in) curtain-sided containers allowed the train to be diverted over virtually any route, including the Settle–Carlisle and Cumbrian Coast lines.

DRS – along with its partners – continues to explore the intermodal market but has not seemed to have much luck with Teesport. During the period 2011–2012, a weekly Malcolm Group train from Elderslie and Grangemouth, and a daily Stobart Group initiative linking Tees Dock with their Widnes distribution centre at Ditton, both foundered. The Ditton trains were a much-publicised high cube container flow using low-floor wagons for P&O Ferrymasters, which was abandoned as uneconomical. Admittedly, I only saw the train half a dozen times, and on all and one of those occasions it was fully loaded, but perhaps I only witnessed the good days. DRS continues to support Teesport and introduced a new connection with Daventry towards the end of 2019.

DRS's first intermodal operation was a trial using piggyback wagons to transport milk from Cumbria to the capital every day for several weeks in June and July of 1997. The railhead was Cricklewood, from where it returned overnight. Carrying a stick-on Milkliner 2000 headboard, the service on 8 July 1997 is taken through the Lune Gorge by locomotives 20303 and 20304.

A pair of the ex-EPS 37s were tried on the Milkliner on 26 July 1997. Locomotives 37609 and 37610 get the milk underway from a wet, windswept Penrith station. The trial was a success, but it was thwarted by the intervention (unrelated to the trial) of the Monopolies and Mergers Commission.

The Minimodal concept was a Strategic Rail Authority Innovation Award winner in 2000. The lightweight cube containers had a W6A gauge classification and could go anywhere on the national network when carried on the Megafret-type intermodal wagon. Yet to receive their Minimodal branding locomotives 33030 and 33025 top and tail the Megafret twin wagon demonstration set at Garsdale on 28 August 2002. (Courtesy Gordon Edgar)

DRS, in the form of locomotive 66420 with coaches for invited dignitaries, supports the MULTIMODAL CUMBRIA initiative at the Port of Workington on 24 June 2011 (locomotive 66302 was on the rear).

Early days for DRS intermodal, 4M30, the 1900 Grangemouth–Daventry passes Wandel, near Abington, on 2 July 2001. The KSA wagons were former Rover Cars cube wagons that had a central hydraulic platform, allowing palletised goods to be loaded and lowered into the well section for increased capacity.

In 2002 DRS received its first Class 66 locomotives, which were crucial to the development of their intermodal portfolio with key logistics players, the Malcolm Group, the Russel Group and Stobart Group. Here, a Daventry–Coatbridge intermodal (partnered with the Russel Group) climbs Grayrigg bank at Hardrigg behind locomotive 66403.

The most notable intermodal in the UK is the daily Daventry–Grangemouth 'Tesco' train. Here, low-emission locomotive 66411, resplendent in Stobart Rail livery, climbs Grayrigg Bank near Beck Houses on 2 February 2007 with 4S43, the 0631 Daventry–Grangemouth.

This picture was taken in the South Lanarkshire hills near Crawford on 24 July 2014 and features 4S49, the 2158 Daventry–Grangemouth. The distinctive Asda and Malcolm containers in the train are clearly seen.

Locomotives 57004 and 57009 on full bore approaching Elvanfoot with 4M82, the DRS 1606 Coatbridge–Daventry service, 14 April 2010.

With 'Malcolm Logistics' clearly displayed on the curtain-sided containers, 4M16, the 0848 Grangemouth–Daventry glides round the curve at Relly Mill, Durham, behind locomotive 66424. The train ran via the ECML on 4 May 2013 due to engineering work on the WCML.

DRS adorned some of its Class 66s with its logistics partners' liveries, namely Stobart and Malcolm. The Malcolm livery and logo was applied to locomotive 66434, seen here at Dilston on the Tyne Valley line with a diverted 4S43. This picture is an example of one reason that DRS seemed to be wedded to diesel locomotives: their flexibility. The train can take a diversion over non-electrified lines. Note also how low the containers sit compared to the locomotives owing to the small-wheeled IKA megafret-type wagons and the 14 m curtain-sided shipping containers, which at 2.59 m high avoid loading gauge issues on most of the network.

The combination of diesel power and low-height containers allowed the 'Tesco' to be diverted via the Cumbrian Coast route following the derailment of a Virgin Pendolino on the WCML at Lambrigg in 2007. Here, on 1 March, locomotive 66407 moves the southbound 'Tesco': 4M48 (the 1540 Grangemouth–Daventry) towards Whitehaven tunnel's northern portal.

In 2014 DRS began to receive its Class 68 locomotives and trialled them over the Shap and Beattock banks in pairs. Locomotives 68022 and 68020 catch the hazy morning sunshine on a freezing 28 December 2016 as they climb past Shap Wells with 4S43, the 0616 Daventry–Mossend.

The second reason that DRS preferred diesels was that some freight terminals were not 'wired' so a separate diesel locomotive was needed to assist where the wires end. The Class 88 bi-mode was the answer, although their introduction was not without problems in poor adhesion conditions. On this occasion though locomotive 88001 takes Shap Bank in its stride as it lifts 4S43, the 0616 Daventry–Mossend 'Tesco' through falling snow at Shap Wells on 19 January 2018.

4M48, the 1854 Mossend–Daventry 'Tesco' emerges from beneath the A74(M) at Stoneyburn, Crawford, on 26 July 2018, powered by locomotive 88007. The return working has historically supported a load from Tesco and other customers, such as Coca Cola.

Tesco's distribution strategy expanded in the south of England and Wales to places like Purfleet and Wentloog in Cardiff, and in Scotland expansion north of the central belt was to Aberdeen and Inverness. Here, locomotive 66422 takes 4D47, the DRS Inverness–Mossend 'Tesco' through Birnam Wood near Dunkeld on 26 June 2019.

Morralee, a typical buff-coloured stone-built farmstead in rural Northumberland, is passed on 4 August 2012 by locomotive 57004 heading the late-running 4E38, the 0232 Ditton–Tees Dock Intermodal, which often ran via Kingmoor and the Tyne Valley on Saturdays to change locomotives.

Chapter 9

Passenger Trains

In 1973 the world's most famous locomotive, *Flying Scotsman*, was the first of many steam engines to visit Sellafield. The plant's extensive internal rail network had two triangles, an arrangement of track on which it was possible to turn a steam locomotive around. In Great Britain's post-steam age railway system, turntables (the historical equipment for turning a locomotive) were just a memory. This made Sellafield the ideal destination for steam train excursions. Travelling up the beautiful Cumbrian coast, visitors could spend time on the 15-in gauge Ravenglass and Eskdale Railway or even visit the Sellafield site itself on the popular free bus tours put on by a management keen to demystify the industry and make it more friendly. The shocking events of 11 September 2001 changed all that. The bus tours were cancelled, and someone decided that bringing a steam locomotive onto the site was a security risk. The additional fences that began to go up around the Sellafield site even swallowed up one of the triangles; the age of innocence was over.

While Sellafield Limited shunned the charter train market, DRS embraced it, once again turning to its Carlisle logistics partner, the Stobart Group. The Stobart Group had been formed after Eddie Stobart was bought out by Westbury Property Fund in a complex city merger involving Hertfordshire Rail Tours. The resulting 'Stobart Pullman' was a first class, silver service, charter train operated by DRS. The train ran circular dining trips and charters to open golf tournaments, Ascot race course and other prestige events in 2008. Apart from the Mk 1 buffet and brake, the Stobart Pullman stock consisted of all first class Mk 3s with a livery based on Stobart's 'S' chevron logo and locomotive 47832 was given a matching makeover. Like some of the freight initiatives it was not the success that was hoped for and was discontinued after less than six months. Locomotive 47832 was to carry a variety of liveries.

In April 2011, DRS secured the contract to run 'Northern Belle' excursions and painted two of its locomotives (47832 and 47790) in a quasi-Pullman livery to match the Pullman stock, with locomotives 57312 and 57305 later getting a similar treatment. By 2018 though, DRS was supplying standard liveried Class 68s to this train to achieve improved performance. DRS has handled all kinds of passenger work in a similar vein to the above, including the Hartlepool Tall Ships event in 2010, and held contracts to work Cruise Saver Travel's 'Ocean Liner Express' trains. DRS also hires out its locomotives and traincrew: in 2012, DRS was awarded the contract to supply 'Thunderbirds' to rescue failed trains on the West Coast Main Line. At that time Virgin West Coast was the main operator until it was replaced in 2019 by Avanti. DRS uses former Virgin Class 57s, equipped with delner couplings and electric train supply, stationing them at strategic WCML locations such as Carlisle, Crewe and Rugby.

The DRS strategy outlines its approach to passenger operations and how it fits in with its 'Nuclear Mission'. Provision Services involves the supply of locomotives, drivers and stock. Wet Lease Services simply involves hiring out locomotives, whereas Charter Services is its involvement with luxury excursion-type trains such as the 'Northern Belle'. When compared to the objectives of maintaining the ability to run a range of trains across the network, taxpayer value for money and retaining the best people, DRS sees the charter market as its best option for passenger trains. Hiring out locomotives seems to be the least favourable option.

On the night of 19–20 November 2009, unprecedented levels of rainfall had culminated in serious flooding across Cumbria, but in Workington and its surrounding area there was devastation. All crossings of the River Derwent west of Cockermouth were so seriously damaged that they were closed or, like the Northside road bridge, were totally destroyed – an event that also resulted in the untimely death of PC Bill Barker. The railway was the only way to cross the river and DRS was tasked with operating an emergency train service. After use on the Stobart Pullman some of the first-class stock received DRS livery and was used on the free shuttle trains that normally comprised three Mk 3 coaches and a Mk 1 brake. Owing to the short journey (just over five miles) they were 'top and tailed' with locomotives drawn from a small pool of classes 37, 47 and 57s. Workington North, an emergency, temporary 'Park and Ride' station, was constructed near Siddick by Network Rail and its contractors in just four days, the whole operation funded by the Department for Transport. The free shuttles ran for six months from 30 November 2009 to 28 May 2010, augmenting existing Northern trains with an hourly service between Workington and Maryport, and calling at the newly built Workington North and Flimby.

The nuclear industry has always had some connection with passenger trains; the station at Sellafield would not exist without the nuclear plant. In the 1960s three trains would transport workers to and from the site from both directions on the Cumbrian coast and, in addition, via an inland route (the former Whitehaven Cleator & Egremont railway). The use of the railway and, indeed, buses by Sellafield workers had declined steadily over the life of the plant, in line with car ownership. Attempts to reduce the number of cars entering the Sellafield site (on environmental and, latterly, security grounds) took a long time to come to fruition and has always been somewhat half-hearted.

A six-week trial started in December 2011, using a Class 37 locomotive and ex-Arriva Mk 2s. At this time DRS provided the footplate crew and guard, with a Northern conductor on board to issue tickets, but nothing came of that trial. Three years later, a second trial did bear fruit, starting on 17 May 2015. This time DRS supplied the locomotives and stock, with some crew involvement early on, but Northern eventually provided the crews.

At around the same time, DRS started to provide a similar set up for Greater Anglia, on the Wherry lines in Norfolk; the 'top and tail' arrangement there persisted right to the end of scheduled locomotive working in September 2019. In Cumbria it only lasted for four months.

From September 2015, DRS provided Mk 2f driving trailers which (in coaching stock nomenclature) are classified Driving Brake Standard Open (DBSO). They are coaches with a driving cab that allows the train to be pushed by the locomotive at the rear. The driver controls the locomotive remotely from the cab situated at the front of the coach. The Class 37 locomotives proved to be unpopular with travellers as they did suffer more failures than their multiple unit predecessors. The fifty-year-old locomotives were probably not suited to the stop–start nature of the Cumbrian Coast route, but there does not appear to be any evidence in the public domain to throw any light on why this may be the case, so speculation is pointless. In terms of failures they were less reliable so were not popular with travellers, even if they did approve of the comfortable Mk 2 coaches. They were certainly very noisy and, when left idling (particularly at Barrow-in-Furness carriage sidings), they had badly smoking exhausts that also generated criticism from residents near the line.

An attempt to improve the situation by utilising a more modern locomotive proved problematic too. DRS had begun to receive its Class 68 locomotives around the time of the start of the locomotive-hauled trains in 2015. The twenty-first century design of the Class 68s was not compatible the DBSOs, which used the old BR blue star multiple working system when operating with Class 37s or the very old Time Division Multiplex system from their early push–pull days. So, the Class 68s had to work 'top and tailed' but the extra locomotive made the trains longer, and it was decided to remove a coach to maintain the previous train length. This occurred because stop boards had been installed at stations so that the driver has an accurate guide to the optimum stopping place. There are a number of reasons for this: for example, it lines up doors with Harrington Humps (a slightly raised part of the platform to assist passengers boarding and alighting) and prevents the tail of a train obstructing level crossings that are located immediately at the platform's end (a feature of several Cumbrian coast stations). As no brake coach was available the DBSOs remained in the train so the trains comprised two standard coaches and three driving vehicles! The Class 68s only operated between March and September 2018 and all Northern Rail's locomotive-powered trains ceased to operate on 28 December 2018.

An early encounter with passenger working for DRS was this rescue of a Class 87-powered Virgin train, 1M08, the 0615 Glasgow–Euston that had failed near Wamphrey, Dumfriesshire, on 6 April 2000. It is seen shortly after being rescued north of Lockerbie behind locomotives 37609 and 37607.

Sellafield's association with passenger trains in the leisure market goes back to 1973 when *Flying Scotsman* was first allowed to turn on the site railway system's triangle, a procedure that was banned by the NDA in 2011 on security grounds. The previous year (on 9 August) having detached from the West Cumbrian, locomotive 48151 awaits entry to the Sellafield complex to turn.

Locomotive 47832, with its Stobart chevron livery matching the stock, heads west along the Tyne Valley line, at Scarrow Hill on 14 June 2008 with the short-lived Stobart Pullman.

On 27 August 2011, locomotive 47832 (along with 47790) now has a livery to match the Pullman-based Northern Belle stock, seen on a Cumbrian coast excursion at Nethertown; two Class 57s also carried this livery.

What remained of the North Side road bridge, five months after the devastating Workington floods on 22 April 2010; downstream, 2T24, the 1030 Workington–Maryport flood relief train is taken across the River Derwent by locomotive 37611.

An immediate problem with Workington North was its popularity: the car park was not big enough! This is evident on 4 December 2009 as locomotive 47790 restarts 2Z25, the 1050 Maryport–Workington service.

After use on the Stobart Pullman, some of the first-class stock received DRS livery and was used on the Workington–Maryport flood relief free shuttles.

Locomotive 47832 restarts the 1050 Maryport–Workington from Workington North on 30 November 2009.

Locomotive 37611 rolls into Flimby on 22 April 2010. The trains were all top and tailed, this one having locomotive 47501 on the rear. The shuttles ran for six months from 30 November 2009 to 28 May 2010 using DRS coaches, locomotives and crew.

The first trial for a Sellafield workers' train started in December 2011 and ran for six weeks using ex-Arriva Mk 2 stock and a single locomotive but was not followed up. Locomotive 37423 sets back the stock from Platform 2 to the Carriage Sidings at Barrow-in-Furness where it will run around, a practice that seems out of favour in the modern railway. At this time DRS provided the footplate crew and guard with a Northern conductor on board.

Another Sellafield workers' trial began in 2015, again featuring the same locomotive. By now locomotive 37423 had received its new winged compass livery and is pictured at Cunning Point north of Parton on 1 May 2015, shepherding a motley collection of stock.

The second trial was followed by the start of the Cumbrian coast service from 17 May 2015. 2C47, Northern's 1731 Barrow-in-Furness–Carlisle, top and tailed by locomotives 37423 and 37688 is seen in the Pow Valley, St Bees on 3 June 2015. There were no DRS-liveried brake coaches available, so this set was using a former Arriva brake.

The second set used on the locomotive-hauled Northern Trains service used a Virgin-liveried brake, seen here in May 2015 at Sowerby Lodge, Ormsgill, between Park South and Barrow. The train is 2C32, Northern's 0515 Carlisle–Barrow-in-Furness, top and tailed by DRS locomotives 37419 and 37611. In the distance, on the left of the picture (across the Duddon Sands) Black Combe broods beneath a bank of cloud.

It was decided to end top and tail operation with the introduction of a DRS driving coach. Here, 5Z60, a Carlisle Kingmoor–Carnforth test train, passes Marsh House, St Bees on 16 July 2015 evaluating a Driving Brake Standard Open (DBSO) coach no. 9705. Sandwiched between the driving trailer and 57011 is locomotive 37401 in fresh large logo BR livery with the Highland Rail stag motif.

The locomotive-hauled sets may not have been popular with all lineside residents and many regular travellers on the Cumbrian coast, but their brake vans were very popular with cyclists. 2C48, Northern's 1156 Carlisle–Lancaster stands at Cark on 12 May 2018 with DBSO 9709 at the front and locomotive 37402 on the rear.

By July 2015, DRS had turned out a Class 37/4 in a former BR large logo livery. The locomotive carried this livery when it was allocated to Eastfield, Glasgow in the 1980s. Restored to its former glory by the Bo'ness Diesel Group, locomotive 37401 is seen here at Nethertown (though at that time missing its white westie dog logo). Note that all the stock, including the brake, is now in matching DRS livery as is locomotive 37402 on the back of the train; however, locomotive 37402 also would soon appear in the large logo livery.

Locomotive 37402 appeared in large logo livery in March 2016 and is seen here on 24 November that year at Green Road, propelling 2C40, Northern's 0842 Carlisle–Barrow-in-Furness. The snow-covered Scafell Pike, Broad Crag and Ill Crag tower over the scene, with Harter Fell looming up on the right of the picture.

The Holly Farm Road Bridge at Reedham, in Norfolk, is a popular place to observe and photograph trains rumbling over the River Yare on the impressive Reedham Swing Bridge. In this view, the swing bridge may be seen through the arches of the road bridge (itself a significant structure) as unbranded locomotive 37422 takes 2J67, the 0747 Lowestoft–Norwich away from the scene and on towards Reedham Junction on 14 July 2016; (locomotive 37405 was at the rear).

Red berries – fast food for wintering birds – are the first signs that autumn has begun. Locomotive 37403 departs Furness Abbey Tunnel on 6 October 2016, heading 2C47, Northern's 1004 Preston–Barrow service.

Locomotive 37424 (alias 37558) *Avro Vulcan XH558* negotiates the severe curves of the Cumbrian Coast Line at Parton whilst heading 2C33, Northern's 0546 Barrow-in-Furness–Carlisle.

Class 68s were drafted in as diesel multiple unit (DMU) cover on The Wherry Lines in July 2016. Pictured on 14 July, 68016 heads 2J73, the 1057 Lowestoft–Norwich (with locomotive 68023 on the rear) over Reedham Swing Bridge as locomotive 156 407, a Norwich–Lowestoft service, heads the other way.

Locomotives 68004 and 68017 top and tail 2C49, Northern's 1140 Barrow-in-Furness–Carlisle, pictured running along the sea wall at Tanyard Bay, Parton on 1 May 2018.

In the Norfolk twilight, 2J88, Greater Anglia's 1900 Norwich–Lowestoft service, formed with DRS Mk 2 stock and top and tailed by locomotives 37425 and 37422, approaches Hadiscoe on 28 August 2015.

Chapter 10

Depots, Locos and Logos

This final chapter looks at the locomotives used by DRS and the livery variations that they carried. Although the company relied heavily on former BR traction, its core fleet was either extensively refurbished in DRS ownership or recently refurbished by previous owners. DRS began operations at Sellafield in February 1995 with only five Class 20 locomotives and a handful of employees. The Carlisle Kingmoor Depot was re-opened by DRS in 1998 and Crewe Gresty Bridge re-opened as a DRS depot in 2007. Ownership of the company passed from BNFL to the Nuclear Decommissioning Authority (NDA) in 2005.

The DRS livery has changed over the years and the photographs in this chapter illustrate the evolution from a plain dark blue livery with basic lettering to a corporate identity in 2012 where its masterbrand logo (a winged compass) was adapted and incorporated into individual locomotive designs.

The year 1997 saw the first six Class 37s arrive in the DRS stable from Eurostar after the collapse of the Nightstar project; three more would arrive in 2002, and there would eventually be a couple of dozen in the stable following purchases from EWS, Harry Needle and other one-off purchases. Although powerful, the Class 37s had route availability '5' so could travel virtually anywhere. The Class 20, 37 and 57s were the largest groups of former BR locomotives, the Class 57 fleet being mainly made up from ex-Freightliner machines and former Virgin Thunderbirds. Four Class 33s were operated between 2000 and 2005, and other classes ranging from Class 08 to Class 87 were trialled but were not deemed successful. The Class 47 and Class 57 locomotives were bought for passenger and fast intermodal freights, although the Class 66 was essential if the company was to compete with existing intermodal operators such as Freightliner, EWS and GBRf.

In 2002, DRS borrowed locomotive 66710 from GBRf to try out on its intermodal operation and subsequently ordered ten of them, leased from Porterbrook. These were returned in 2008 in favour of versions that met Stage IIIa of the European Directive on Emissions and settled on a fleet of approximately twenty-four, reducing to about twenty as the Class 68s came into the fleet. However, the Class 66 could not meet Stage IIIb of the Directive, which came into effect in January 2015, so any new locomotives procured had to be sourced elsewhere.

The initial batch of Class 68s were built by Vossloh Espana, which became part of the Swiss firm, Stadler, to a design derived from the Stadler Eurolight. Powered by a sixteen-cylinder 3,800 hp Caterpillar engine, they were to be leased from Beacon Rail and a total of thirty-four was ordered. Most of them, however, were sub-leased to passenger train operators Scotrail, Chiltern and Transpennine.

The Class 88 locomotives are described as a proper bi-mode locomotive, they have a Bo-Bo wheel arrangement and, though they are more powerful than older British four-axle AC electric locomotives, they still only have eight wheels to transmit that power to the rail. However, it is claimed that superior computerised traction management makes up for that deficiency. Equipped with pantographs to collect

25 kV overhead they also have a 950 hp Caterpillar diesel engine that was installed to do the so-called 'last mile', to take a heavy electric freight into a non-electrified yard or terminal. So, one might think that the diesels would not be geared for running at line speed over distance, but in fact they seem to be excellent at hauling lighter trains to schedule as well as being able to take juice to haul a heavy freight over the Shap and Beattock banks. With regard to the flask trains, if you are forced by policy to have two locomotives on every train then it makes sense to use low-powered diesels, so the Class 88 works very well for DRS.

Now for the counter argument: should the taxpayer be paying for expensive, electric 100 mph mixed-traffic or freight locomotives to tootle around secondary lines with short light freights? It has to be said, too, that these limited observations must be set against professional drivers' reports: they have experienced problems first-hand on the northern banks, where pairs of Bo-Bo or single locomotives with Co-Co wheel arrangements have traditionally done the job. Even in their full electric mode, slipping and problems with sanding have been reported on the banks in poor adhesion conditions, although DRS states that better driver training and equipment modifications solved those issues. All things being equal, the small diesel engine in the Class 88 copes very well – but all things are not always equal. For example, the Mossend Freight Terminal, serving the Peter D Stirling logistics operation, is a steep branch and with a fully loaded train, a Class 88 in poor rail adhesion conditions has struggled to lift a train back up into the North Departure line. A multi-tasking gem or a 'Jack of all trades and master of none'? Time will tell.

The depot where DRS began: Sellafield on 17 December 1996, with two of their first locomotives (20301 and 20302) during a brake test prior to running a tanker train to Ince.

The Kosovo 'Train for Life' carried aid from the UK to Kosovo in September 1999, in support of the United Nations Kosovo Force (K-FOR), a peace-keeping initiative during the conflict following the break-up of Yugoslavia. The train was powered by DRS locomotives 20901, 20902 and 20903, with drivers from 79 Railway Squadron (Royal Corps of Transport). Here – after their sojourn abroad – two of the engines, 20901 and 20902, are back in the UK under maintenance at DRS Carlisle Kingmoor on 15 April 2000.

Kingmoor Depot was re-opened by DRS in 1998 and is able to carry out major maintenance tasks, such as bogie and traction motor swaps. Locomotive 37515 is nearest the camera and, on the left, locomotive 37609 is being separated from its bogies on Matterson 35-ton jacks.

On shed at Carlisle Kingmoor on 5 November 2006 are locomotives 08834 and 08892, both carrying DRS lettering. DRS considered using them on the Sellafield–Drigg trip but they saw little use, eventually being sold to Harry Needle in 2009.

Pictured alongside locomotives 20905 and 37609 outside the DRS depot at Carlisle Kingmoor on 15 April 2000, this former army, Barclay-designed, diesel mechanical 0-4-0 was trialled by DRS as a depot shunter but eventually moved on to the Eden Valley Railway via Haig mining museum. Having a yellow (albeit inverted) 'V' on a blue livery (and perhaps the fact that it was built by the same Vulcan Foundry that built the prototype deltic) seems to have been enough to earn this little shunter the ironic nickname 'DRS Deltic'.

The first-phase DRS blue livery, with light blue blocks and simple lettering, modelled by Class 20s entering Sellafield with a flask train from Barrow Ramsden Dock in June 1999. The lead locomotive in this picture (one of the second batch of Class 20s purchased in 1997) is former Pete Waterman-liveried locomotive 20312, which still has a black roof. Locomotive 20305 behind it (one of the original batch) has a light blue/grey roof.

Kingmoor depot in February 2003. Just beyond the left-hand overhead line equipment (OHLE) stanchion, the fenceposts may be seen placed loosely in their holes, ready for the security fence to go up. The first-phase DRS blue livery is applied to locomotive 33025, enhanced with minimodal branding, seen passing the depot with tanks from Carlisle Yard to Sellafield.

Locomotive 37610, still with Channel Tunnel cast metal roundels, and locomotive 37609, with scars from the removal of its roundels, stand at Penrith on 26 July 1997. They still carry grey EPS livery, overlaid with phase one-style DRS vinyls. Rather than cyan lettering and light blue blocks, the vinyls have white lettering and blocks on a dark blue base. Both locomotives also carry white Tankfreight branding at central left bodyside.

Locomotives 37079 and 37194 rumble over the South Tyne viaduct, Ridley Hall, on 17 July 2006, with 6M60, the Seaton–Sellafield flasks, showing the first-phase DRS blue livery as applied to Class 37s. In the twenty-first century the company web address was inserted into the light blue block below the main DRS lettering. Locomotive 37029's light blue blocks have not weathered very well.

First-phase livery on locomotives 37610 and 37607 as they pass through Annan station on 17 February 2000. This picture is included to record the DRS foray into coal trains, owing to EWS locomotive and crew shortages. This is 6M40, the 0706 Ayr–Fiddlers Ferry MGR, which DRS worked as far as Carlisle.

Sandite-fitted locomotive 37229 (left) and locomotive 37029 on shed at Carlisle Kingmoor, 15 February 2003. The dark blue first-phase livery may be seen on locomotive 37029 (on the right of the picture) with cyan lettering and the light blue lines and blocks. However, the difficulties of fitting the DRS branding on a Class 37 bodyside is also evident. The Sandite port has been inserted into the design space taking out part of the blue line and a small part of the lettering. The bottom right-hand corner of the blue 'website' block has been taken out by one of the sandbox fillers, although this could have been avoided by locating the whole motif slightly to the left.

Another freight that DRS locomotives were utilised on for several years was 6J37 the 1251 Carlisle Yard–Chirk Timber Working. Locomotives 57008 and 57002 lean into the curve at Redhills, Penrith, on 3 November 2009, adorned with the second-phase DRS dark blue livery. The Direct Rail Services legend replaces the abbreviated version on a much larger blue block that is connected to the new green compass logo by four stripes with blue to green colour graduation. Larger website lettering completed the transformation, which looked impressive on the broad flat bodyside of the Class 57 or 47.

The second-phase livery did not really fit the Class 20s: the four stripes were omitted, leaving the compass on the pannier tank disconnected from the lettering. Pictured on 7 June 2013, 6C51, the 1258 Sellafield–Heysham flask train, approaches the Oxcliffe Road Bridge between Morecambe and Heysham with locomotive 20309 leading and locomotive 20305 in the rear.

The phase two livery did not fit well on the Class 66 either. The blue block was a lighter blue/grey but was adulterated by the large bodyside grille and the Direct Rail Services legend was much smaller. On locomotives like this one with mid-bodyside doors the compass logo was also behind the door handrails. Here, 4S43, the 0631 Daventry–Grangemouth 'Tesco Express' is powered through Beck Foot on 22 August 2007 by locomotive 66415.

The third phase of DRS corporate identity centred on a winged compass logo, with the Direct Rail Services legend appearing in lower case with initial capital letters. This has proved to be much more adaptable to a variety of locomotives and sits very well on locomotive 66432, seen at Barrow Docks working 6C31, the 1840 aggregates train to Sellafield on 21 February 2018.

The winged compass identity is adaptable. A green bodyside graduated vinyl provides a foundation for the logo on Class 37s, demonstrated here by heavyweight locomotives 37716 and 37609 as they top and tail the 1258 Sellafield–Heysham flasks, passing Limestone Hall crossing on 8 March 2017.

The winged compass on DRS 'Thunderbird' locomotive 57307 is seen at Preston on 10 March 2014, along with a message to deter would-be cable thieves and the logos of supporting organisations. When DRS took possession of the Class 57s from Virgin it made a £5,000 donation to The Railway Children charity to maintain the use of the *Lady Penelope* nameplate, the only one of the nameplates featuring ITV *Thunderbird* characters to survive into DRS, as shown in the next photograph.

A demonstration by a Cumbrian firm ast transport branding is taking place at Kingmoor Depot on 18 July 2015, during the open day to celebrate 20 years of DRS. Locomotive 57307 is being transformed from DRS blue, formerly adorned with 'Cable Thieves: We're Watching You' lettering to the latest vinyl compass identity with '20 years of Direct Rail Services' branding.

The cabs of two Thunderbirds, locomotive 57304 and 57309, at shift changeover in Carlisle on 13 June 2019. The smaller winged compass logos are displayed with the word 'Direct' in a larger-size typeface.

The Class 68s arrive. Locomotives 68016 and 68017 were unloaded at the port of Workington on 26 October 2015 in a plain blue paint scheme. Here, the first locomotive is removed from the hold of Netherlands-registered cargo vessel *Douwe-S*, which had sailed from Sagunto, near Valencia, in eastern Spain.

Some Class 68s, especially those destined for passenger operators, received a basic winged compass logo on the blue base. Here, locomotive 68030 heads 6C51, the 1258 Sellafield–Heysham flask train, passing frozen floodwater in fields near Limestone Hall Crossing on 1 December 2017. Locomotive 68018, on the rear, has full body-length green vinyls with compass livery.

The standard livery for the Class 68 locomotives is a full-length abstract, dynamic green vinyl, superimposed on the dark blue base. The white-on-green compass is augmented by four speed-blurred white stripes. The 'masterbrand' winged compasses only appear on the cab sides. Locomotive 68002 backs wagons from the 0605 Sellafield–Barrow Marine Terminal into its destination on 6 June 2018; (locomotive 37423 may be seen inside the terminal).

When viewed in close-up, a map of Cumbria is revealed in the white parts of the compass.

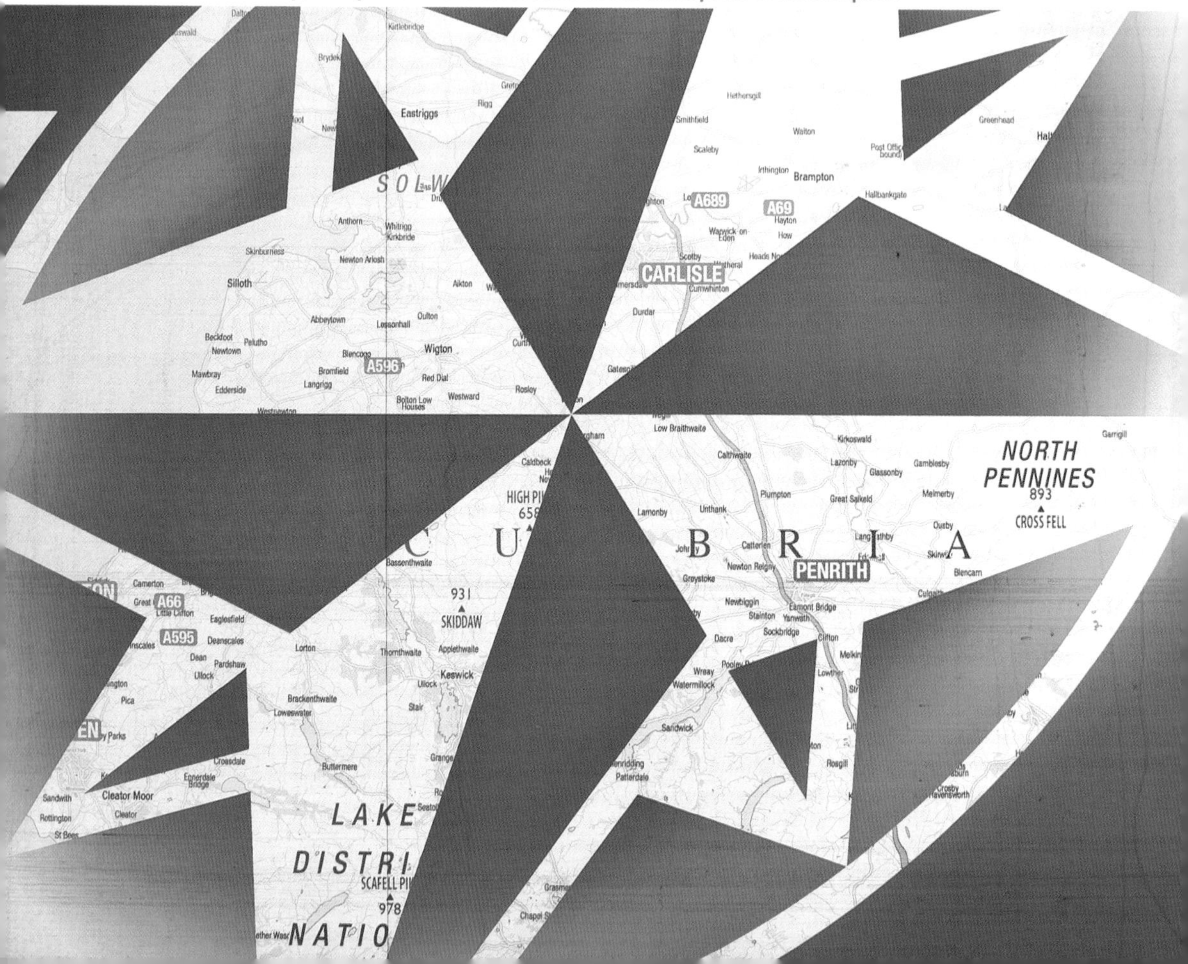

Inch by inch, locomotive 88002 descends the ramp from the road trailer to the rails at Brunthill on 24 January 2017. The locomotive had arrived at Southampton Docks and was moved by road to Brunthill Freight Depot in Kingstown, Carlisle. The Class 88s received the same compass logo (with four speed-blurred stripes) as the Class 68s, but the blue base colour was augmented by a corset of lines intended to convey the impression of a network of electric wires.

A close-up of the No. 1 end of locomotive 88005 after arriving in the country at Workington on 1 March 2017. The pantograph may be seen, neatly tucked into the roof profile, and the striking electric livery is shown to good effect with the red representing electrical power coursing through the wires. Finally, the locomotive's name, *Minerva*. Eight of the batch of ten locomotives take their names from classical mythology in homage to British Railways early 1500 V DC electric locomotives of class EM2 and an EM1.

The world into which DRS was born: the stunning coastal scenery and mechanical signalling of the Cumbrian coast. In 2019, DRS signalled its intention to run down its heritage fleet so images like this, of locomotives 20901, 37229, 33025 and 33207 at Parton in August 2003, will be just a memory. It is often remarked that on DRS trains the locomotives outnumber the wagons; in this picture the locomotive *types* outnumber the wagons!